50 SCIENCE *Blessings*

5×10^1

SCIENCE

Blessings

PRAYERS, TOASTS, & PROVERBS *for*
Holidays, Special Occasions, *and* Daily Life

J.G. KEMP

CONTENTS

Introduction: On Science Blessings and
Dinosaur-killing Asteroids   9

Traditional Holidays

Thanksgiving / Chicxulub Day   18

Christmas Eve   22

New Year's Eve   26

Valentine's Day   29

Mother's Day   31

Memorial Day   33

Father's Day   36

Independence Day   39

Celestial, Math & Science Holidays

Winter or Summer Solstice   44

Full Moon   47

Mole Day   48

Pi Day   51

Special Occasions

Birthdays   56

Weddings   58

Anniversaries   62

Graduations   64

Daily Life

Mealtimes   68

Mornings   74

Evenings   78

Blessings and Goodbyes   80

Coffee, Tea, or Alcohol Time   83

Science Proverbs   86

An Author's Request

To my mother and father, who placed me on the tree of life, and who gave to me both ancient and modern wisdom. I humbly thank you.

—J.G. Kemp

Introduction

On SCIENCE BLESSINGS
and DINOSAUR-KILLING ASTEROIDS

The idea for this book of science blessings
first occurred to me four years ago, shortly
before Thanksgiving. I was teaching Earth
science to a class of spirited fourteen year
olds, and we were discussing the discovery
of Chicxulub crater, the massive crater in
the Yucatán made by the six mile wide
asteroid responsible for the killing of the
dinosaurs — or so the theory goes. This
asteroid was very important to us, I
explained, because the sudden demise of
the dinosaurs set the trajectory of life on a
new course, a course which, 65 million
years later, has led to us, *Homo sapiens*. And

since this event was far more important than say, the survival of a small colony of pilgrims in New England four hundred years ago, I proposed that in addition to our Thanksgiving celebrations we should also celebrate that asteroid impact, and perhaps call it: Chicxulub Day.

This suggestion was met with a few laughs at the time, but later I thought: Why not? Why not celebrate Chicxulub Day? And why stop at one holiday? Why not have an entire year of science-themed holidays, and a book of blessings which could be read aloud to remind celebrants of the holidays's importance?

Well, four years later, I am happy to say that, although I haven't invented an entire year's worth of science-themed holidays, I have written this collection of science blessings for traditional holidays, as well as a few extra prayers, toasts, and proverbs for other occasions.

Here you may ask, "What is a science blessing?" or perhaps, "How do you turn a prayer, which in most cases is a plea to God or a supernatural power, into a prayer of science, which makes no claim of super-nature?"

For the purpose of this book, I have defined a science blessing or prayer as: *A sentiment, typically of gratitude or thankfulness, paired with the facts and conclusions of science, with no reference to that which lies outside the realm of science, such as God or the supernatural.*

Here some will argue: "But how can you remove God from a blessing?" or "What's the point of praying if you're not praying to anything?" To these objections I would simply reply: that gratitude is universal, and although it is natural, when we are grateful, to want to thank someone, one can appreciate a gift without knowing of a giver; one can be grateful for creation without referring to a creator, grateful for the universe without contemplating a universe-maker; that it is sufficient and good to appreciate a gift itself, creation itself, nature itself, life itself, the universe itself, nothing more. And so this is the purpose of a science blessing—to express gratitude for the universe itself, for reality, as it is, without the inclusion of that which science can make no claim.

Some might then argue that these blessings are therefore godless, but that is not so. In their unaltered form, these blessings *are* especially intended for people with no particular faith—perhaps those science-minded atheists or agnostics who wish to express their gratitude without com-promising their reason, or who wish to pray but find it difficult to pray *to* something. However, because these blessings make no claim or reference to any particular faith system, they can be readily adapted by people of any faith. For instance, if you wish to bracket these prayers with phrases typical to your faith, such as "Dear God", etc., then feel free to do so. That being said, however, I must inform you that these blessings *do* contain the established facts and conclusions of mainstream science, such as the fact of evolution and the conclusion that our universe is very very old (over thirteen billion years). If your faith system is at odds with, or denies, these firmly established conclusions, then I can give you no advice, and perhaps this book is not for you.

You may notice a few reoccurring themes in these blessings, themes which I consider the most wonderful in all of science. The first, is that matter, made of atoms, can be traced back through evolutionary and deep time to the most extraordinary situations, such as exploding stars (like the often stated "we are made of stardust" line); and the second, is that we can do the same thing with energy—that energy is conserved and "changes forms" and flows from the sun to plants to your thoughts, for example. These findings bring the deepest past into the present and unify the phenomena of the cosmos into a glorious web of interconnectedness.

You will also notice that I have retained the traditional use of the word "amen" to conclude many of the prayers. Amen is often translated as "so be it", and is perfectly acceptable for a science blessing. It is an ending ingrained in my concept of prayer, and it needn't be seen as a declaration of faith or a reference to any particular religion. Again, remove it if you would like, or replace it with a word that feels more natural to you.

I imagine that these blessings will make some people of faith feel uncomfortable. I would only try to reassure them that whether someone has faith or has none, or whether your God, your neighbor's Gods, or no god listens to our prayers, we can still be grateful together, and that we are still brothers and sisters; we can share the same table, the same meal, the same celebration.

Finally, it is my ardent hope that you find in these pages something which resonates deeply with you, something which strikes a chord, something which touches upon that beauty which words can never fully express. I hope that these most wonderful insights from science may touch your heart, and remind you how inexplicably glorious it is to be aware in this universe, how miraculous to simply be alive and breathing!

.

Traditional Holidays

A Prayer for Thanksgiving Feast

A Prayer for Christmas Eve

A Toast for New Year's Eve

A Toast for Valentine's Day

A Blessing for Mother's Day

A Prayer for Memorial Day

A Blessing for Father's Day

A Toast for Independence Day

A Prayer for Thanksgiving Feast
(Chicxulub Day)

1

My friends,
Look you at this feast before us. Grown of
earth and sun. Diversity and subtlety of
colors. Wafting and mingling of smells.
A medley of textures. A swath of
temperatures — trembling atoms in chaos.
A great cornucopia of chemistry. The
drama of life, built by cells under the
direction of genes and the immutable laws
of physics.

My friends, look you at this feast before us.
Life, which gives *us* life. Matter, which
becomes *our* matter. Energy, stored —

but for a moment—before flowing into us
and through us and out of us. We eat to
live another day, to breathe another breath.
How marvelous this play! How glorious
this grand show of life of which we
partake!—the very stuff of our thoughts,
our actions, our dreams!

My friends, look you at this feast before us.
Imagine the days here, the countless
moments: photons captured and stored—
joule by joule, calorie by calorie, electron
volt by electron volt—one by one, day by
day. Imagine the sprouting of seeds, the
showers of spring, the cool rumbling
thunder, the crisp cloudless nights, the
fields that bow to a warm breeze as
crickets chirp… chirp… chirp…
Sunrise, sunset, sunrise, sunset.

Imagine the journeys here — of water, of
minerals — atom by atom.
From ocean to cloud to stream. From
magma to rock to dust. From soil to stem
to leaf. Eons of time, converging now.

My friends, this feast before us comes from
earth and sun, yes, but that is not all.
Indeed, it is the heritage of our human
family. These very plants, season by
season, seed by seed, kept alive and
improved by generations of farmers. Sown
and reaped, harvest by harvest, with tool
and plow, the legacy of invention. And
brought to us, by many people, from many
places, on many paths, from farm to table.

My friends, we are all connected.
Everything is connected. Through time
and space. To the universe,
and to each other.

And so for this, for the great bounty of the
cosmos; and for the parts that each of us
play, in this great human endeavor, (and
for the dinosaur-killing asteroid that struck
our planet 65 million years ago)*, we are
forever thankful.

Amen.

*See introduction. Omit if you would like the tone of the blessing to remain tranquil. Include if you wish to end the blessing with an impact, and a guaranteed conversation starter.

A Prayer on Christmas Eve

2

My friends,

Tonight we celebrate the Christmas story:

The story of a baby, the struggle and

triumph of a family. In the darkness, in the

cold, in a stable, among animals, a man

brings his laboring wife. The woman,

away from home, away from help, gives

birth to a child. The child, his first breath,

crying, nursing. His mother, exhausted.

His father, anxious.

The smell of hay (wafting volatile

compounds from plant decay), the flicker

of candle light (perhaps burning oil,

pressed from last seasons olives, or wax,

built-up in hexagonal form, bee by bee,
from flowers and field), the breath of a
cow, the smell of manure (thriving with
microbes), dirt and dust (worn by ancient
weather and scattered by eons of wind).

All there, in the stable.

My friends, this is Christmas Eve! A
perfectly mundane and gloriously ordinary
night! The simple and incorruptible purity
of a lonely stable, away from the city, away
from society, away from gossip and politics,
from condemnation and prejudice. A clear
midnight silence. A family, new and whole,
on a journey; in and out of sleep; mother
and father and child; and the Earth turns
and the morning comes and the days pass.

My friends, we gather on Christmas to celebrate the birth of a man — a man who, years later, would preach tolerance, acceptance, forgiveness, brotherhood, communion.

My friends, science has shown us that, when we determine the molecules of air in a single breath and the number of breaths in the atmosphere, it is likely that every breath we breathe contains atoms that passed in and out of the lungs of that child on the first Christmas night.* In and out of that family, and of the cow in the stable, and of the sheep in the fields, and of the lungs of everyone living on that day, and the day before, and the day after; and is not that too a message of brotherhood, of communion?

My friends, tonight we are thankful, and
amazed, and humbled: by the miracle and
magic that is life, birth, family, and indeed,
the whole world.

Peace on Earth and goodwill towards all!
Merry Christmas!

*this calculation is often made in reference to "Caesar's
dying breath", and it is probable (depending on your
assumptions), that there is a molecule in every one of your
breaths that was also in Caesar's last. An internet search
will give you many fine examples of this calculation, as well
as discussion of the assumptions involved.

A Toast on New Year's Eve

3

My friends,
Tonight we celebrate time — that unfeeling
flow of events, from one to another, which
has led us here... to this day... this hour...
this minute... this second...
Ever moving. Uncaring. Endless.

Another year has passed, like the year
before, and the year before, and a new year
lies waiting, shrouded in hope and worry,
possibility and potential.
Yet we are here, at this moment, a sliver in
time embraced by the eons before and
behind us. Now. Among all.
We are here now! And the past is lost.

It is a curse to know this—the loss of time.

Tonight we bring to mind the faces of those
we will never see again. We remember the
year's joys and sorrows, the good times and
the bad, yet we remember them less and
less, as time marches on. Relentless,
Uncaring. Endless.

But my friends, look up! Tonight our way
is forward, towards the future, the hope of
a new year, the work to be done, the days
to be filled, the seconds to be savored.
The past is gone, yes, but the future is
before us!

Tonight we celebrate time.
We cry and we laugh.
We cry and we laugh.

Remember...

remember...

remember...

and forget.

Time. Giver and taker. Master of us all.

My friends, raise a glass!

Kiss your loved ones!

Happy New Year!

A Toast for Valentine's Day

4

The science of romantic love is this:
because nature selects bodies, thoughts,
and feelings that survive and reproduce, I
am endowed—the gift of millennia—with
an elixir of neurochemicals: hormones
which, when I see or smell an attractive
mate, incite feelings of desire and lust. My
pulse quickens, my skin sweats, my mind
whirls, my intentions are clear—an animal
whose urge is for pleasure, to reproduce,
quickly and often.

Does knowing this diminish love? Is our
romance a mere trick, played by selfish
genes? Are we but tools for their survival?

It is no matter. For whatever the case may
be. Your eyes are heaven. Your hair is
silk. Your skin, velvet. Your neck is a
target. Your smell—ecstasy! Take my
dopamine. Control my epinephrine. My
vasopressin and oxytocin are yours and at
your command!

Be *now* my valentine…

A Blessing for Mother's Day

5

What can I say, my mother? You give the
years of your life, and what do you get in
return? You carried me, birthed me,
cradled me, nursed me, cleaned me, fed me,
dressed me, taught me, warned me, scolded
me, served me, encouraged me, praised me,
hoped for me, cried for me, worried for me,
prayed for me. Yet what have you received
for these things?

My mother, you have received little, for all
these years, the best of your life, you have
been digging a hole. A hole in the ground.
Deeper and deeper, into the dark.

You have done this without recognition,
without appreciation, without recompense.

But mother, you have forgotten. You have
been digging a hole, but that hole is a well!

Day after day you dig, and often alone, but
from those depths you bring forth life.
Life to replenish the lives of your children.

Today, Mother, I remind you,
and I remember you.
And to honor you I work alongside you,
where you so humbly dig,
the well for your family.

Happy Mother's Day

A Prayer on Memorial Day

6

On Memorial Day we remember those
who made the ultimate sacrifice, who paid
the ultimate price, who gave their lives, so
that we might live.

For these brave fallen we are undoubtably
grateful, but I ask you, how do we honor
that sacrifice, how do we repay our debt?
One day of remembrance? A few solemn
words of thanksgiving?

On the battlefield of Gettysburg, Abraham
Lincoln addressed*, "that we here highly
resolve that these dead shall not have died
in vain."

So I ask you, have those whom we remember today died in vain? What have we resolved to make us worthy of their sacrifice?

I say the greatest honor to those fallen is to live better lives ourselves: to add value to the world, to learn more, to love more, to give more, to be more. To commit ourselves to the pursuit of knowledge and discovery that improves our collective condition, and expands our collective freedom.

Charles Darwin said, "A man who dares to waste one hour of time has not discovered the value of life."

We have life. We have time. We have it, in
part, because of those who died in our
defense. And so today we remember their
sacrifice, and we commit, while our hearts
beat and our lungs respire, to waste no
more time, to live our lives to the best and
the fullest, to value our brief hour, for all
too soon we will join those fallen before us,
in never ending death.

*the word "address" for the Gettysburg Address could not
have been better chosen, as address means both "a formal
speech" and "to appeal, request, or plea". It is the second
meaning that I emphasize here.

A Blessing for Father's Day

7

A father grizzly bear, if he is hungry
enough, will thoughtlessly devour his
children. A father emperor penguin will
stand, for months, in the freezing dark
antarctic, cradling his egg between his feet.
A father lion will let his children starve, to
death, rather than go without food himself;
yet a father hardhead catfish safely carries
his eggs in his mouth, eating nothing until
they hatch and swim away.

But lo, we are not bears, or penguins, or
lions, or catfish; we are humans. And it is
not your child eating or egg balancing
habits that are my birthright, it is your

wisdom, for you are my father, and we are

Homo sapiens — wise man.

Father, you have given to me both the

wisdom of your father, and wisdom you

yourself have discovered.

(The wisdom to care for my body,

mind, and spirit.

The wisdom to listen, in rapture, to music.

The wisdom to learn, always.

The wisdom to seek happiness.

The wisdom to be kind, honest,

and dependable.

The wisdom to seek solace in the woods.

The wisdom to wake, refreshed, under the

blue dome of heaven.

The wisdom to climb mountains, to set

goals, to pursue excellence.

The wisdom to wonder.

The wisdom to demand evidence.)*

These things are my birthright.
And father, I am honored to carry the
baton of your wisdom. I am grateful to be
linked beside you in the chain of our
species. And I consider it a great privilege,
that among all, whether bear or penguin,
lion or man, I may call you: *"my* father"

Happy Father's Day

*This wisdom I have received from my father. Please feel
free to revise it with wisdom received from yours.

A Toast for Independence Day

(This toast is particularly effective when read
immediately after the Declaration of Independence)

8

Ah! Exploding fireworks day!
Pyrotechnics day! The day we celebrate
the independence of our country by
blowing stuff up! And gawking at the
showers of flaming embers in the sky!

722 years before the signing of the
Declaration of Independence, in the year
1054, a star appeared in the sky — bright
enough to see in daylight for nearly a
month, and visible at night for nearly two
years.* We have learned since that this
light was a supernova, a stellar explosion,

the death of a massive star, only 6.5 light
years away; and the remnant of this
supernova is the much studied and
hauntingly beautiful crab nebula, that
mysterious turquoise web of vermillion
tendril and crimson filament.

Ah! What an explosion! More energy
released in seconds than in our
sun's entire life!

We have learned through science that
stars, and supernovae in particular, are the
forgers of atoms, the builders of elements;
and that in the course of these events, our
planet, a new world, was formed from rich
elemental ash, the literal star dust
of eons ago.

So today I propose a toast, not just to the
independence of our country, but to our
independence from the crushing, seething
plasma of bygone super-massive stars. I
propose a toast to our dissolution from
their gravitational tyranny, to our
liberation — cast off in a display of immense
stellar fireworks.

My friends, I propose this toast because
this truth is by no means self evident; it was
a secret which had to be gleaned from the
mysterious universe by the tireless
searchlight of science! Science, that noble
endeavor from which we truly assume the
powers of the earth!

And so, to the potassium in your hot dog,
the oxygen in your lemonade,
your aluminum can,

your flag of carbon thread,

the instruments of brass: copper and zinc,

the showers of flaming calcium,

the fires of magnesium,

the cannons of iron,

and the rockets's red strontium glare!

To life, liberty, and the pursuit of science!

Happy Independence Day!

*a case has been made, from analysis of Chinese texts in
particular, that this star appeared on July 4th, 1054.

Celestial, Math & Science Holidays

A Toast for Winter or Summer Solstice

A Prayer on a Full Moon

A Toast for Mole Day

A Toast for Pi Day

A Toast for Winter (or Summer) Solstice

(for summer solstice, change instances of "south" to
"north" and instances of "north" to "south")

9

For the past six months, by degrees, day
after day, the arc-line traced by the sun has
been shifting (south)ward in the sky. And
today, that path appears to reverse its
(south)erly trend and begins to move
(north)ward again.

This solstice day gives us time to pause and
to reconsider the facts. That the Sun does
not move over us, but that we spin around
it. That the Earth is not flat, but a sphere.
That we hang, unfixed, in the blackness of

space —dig below and reach the other side,
and then emptiness, in all directions;
empty, but for a rare and unlikely ball of
wandering matter.
And yet, we are children born on a great
flat Earth, and a charioteer drives the Sun.

On this solstice day we confess our
stubborn attachment to our child-like
perspective. It is hard to shift. It takes
effort. Our imagination is feeble to
visualize the grand geometry of the
cosmos, the shapes and relations in the
heavens. We are but dust on celestial
clockwork.

And so today, as the (north) axis of our
spinning ball home reaches that angle
which is greatest when pointed away from
the imaginary fixed vertical axis of the Sun,

which we define is perpendicular to the

plane of Earth's orbit, we celebrate.

To Earth and Sun and Science!

Happy Solstice!

Prayer on a Full Moon

10

Oh mysterious sibling. Why should mass
be attracted so? Why? Why!? Were you
but closer, or larger, or slower or smaller,
or faster or farther away.
Your orbit oblonged or inclined.

Birthed from destruction. Held fast by
force that congeals (to a point) this
spherical dust, and joins in dervish dance.

Marble of blue and marble of gray, round
the shining and shimmering face that
bathes us. Lovely sibling, companion, our
friend in the dark.

A Toast for Mole Day

(6:02 am to 6:02 pm on 10/23)

11

Please forgive the rough estimates.

One mole is defined as the quantity of
atoms in 12 grams of Carbon-12. Now, 12
grams of carbon is a 60 carat diamond, and
a 60 carat diamond would be about the size
of a quarter on your finger. That's a big
diamond! That's 12 grams of carbon; that's
a mole of atoms.

Six hundred two followed by twenty one
zeros. Six hundred two sextillion.
Six point zero two times ten to the twenty
third!

What a triumph of mathematics and
science! To decode the atomic nature of
reality using experiment, logic, and the
magic of numbers.

Atoms are unimaginably small, and there is
an unimaginable amount of them. I mean
that quite literally. Our minds simply
cannot imagine the size and quantity of
atoms around us. But we can calculate
them, *oh yes*, and represent them, and build
with them! As small as they are, they
could not stay hidden forever!

Today is mole day! A celebration of
chemistry! A celebration of chemists, who
have dug through the mixed matter of the
cosmos and brought forth her separate
parts; who have quantified the

unimaginably small with astounding
precision and accuracy!

Today we celebrate the Central Science!
The Science to which all other science
must humbly bow and pay homage!

To chemistry! To atoms! To the mole!

A Toast on Pi Day
(3/14)

12

Do you remember the first time you
tasted pi? The first time you sensed the
flavor of its mystery? It is a taste that
eludes many, and it is often an acquired
taste, but it is a taste which, once noticed,
lingers long after, brûléed in memory, a
mental dessert to be savored,
again and again.

Perhaps pi was your first taste of infinity,
of irrationality, of transcendence, your first
taste of a profound simplicity in
mathematics, in which the simplest of
relations can be the most profound.

A circumference and the longest line across
it. That is all. Yet ruminate upon it!

This is why we love pi: It is thrilling. It
makes us wonder. It gives us delicious
hope that we, intelligent apes, are capable
of great understanding.

Yet pi humbles us. Pi is something so
basic, so elementary in math, which defies
our prejudice for integers—our instinct to
merely count things. Pi reminds us,
perhaps, how unequipped we are to grasp
the full mystery of the cosmos. A glimpse,
perhaps, a flash in the dark, a brief taste,
held for a delicious moment and then
swallowed by the sea of thought that is:
living, eating, surviving.

In one bite pi gives us simplicity,
profundity, infinity, humility.
Taste and aftertaste.

What a fine day. To celebrate such a
number by gathering with friends and
eating pie. What a fine day indeed!
Thank you.

Special Occasions

A Toast for Birthdays

A Toast for Weddings

A Toast for Anniversaries

A Blessing for Graduations

A Toast for Birthdays

13

Today we celebrate your birthday. When
the blood coursing through your infant
lungs began to take up oxygen from the air.
When your skin first felt the cold chill of
evaporation, your eyes the first light
through your pupils. We celebrate the day
your inner tube was colonized by bacteria,
comrades in digestion. We celebrate the
day your newborn brain learned of hunger,
taste, and satiety; loneliness and comfort;
abandon and safety.

But that is not all. No. Today we celebrate
the very universe as it is, because of your
life—the impact you have had on the

world. Today remembers the birth of
worry and hope in your parents, the birth
of concern and approval in your relatives,
the birth of countless new thoughts and
new feelings — in the minds and hearts of
those who know you and love you.

But most of all we celebrate the ripples of
action that have spread from you and have
changed our chaotic world in countless
ways we can never know.

Today is your birthday. You did not
choose to be born. You have shaped us, as
we have shaped you. You are here, and we
are glad!
Happy Birthday!

A Toast for Weddings

14

There are many symbols for marriage:
interlocking rings, interwoven hearts,
knots tied, hands held. Today, I would like
to propose two more.

The first, is two arrows pointing towards
each other. Now, this symbol, of course,
represents a convergent boundary of
Earth's continental tectonic plates; two
continents—forced together; slabs of crust,
dragged along by dense, molten, seething
magma—colliding as one.

Now, at convergent boundaries mountains
are uplifted tens of thousands of feet,

rock from deep underground is thrust up
to the surface in towering and spectacular
peaks and crags; the most magnificent
features on our planet are born.

But it happens slowly, so slowly.
Millimeters a year. Unnoticed. And so too
does a marriage grow. You are brought
together by forces, often unseen yet
powerful, and your love will grow—slowly,
so slowly—as you live and age and build
each other up—through wind and ice and
rain—and over time your marriage is a
majestic and triumphant mountain, made
of rock. Two arrows, pointing towards
each other.

The second symbol that I would like to
propose is this: two circles, also with
arrows pointing towards each other.

Now, this symbol, of course, represents the thermonuclear fusion of hydrogen isotopes under the immense heat and pressure in the cores of stars. Two atoms of hydrogen, their repulsive nature overcome, merge to become one, an atom of helium. Now, by itself, hydrogen is reactive, unsettled, always looking to give and take, dissatisfied, as far as elements are concerned. Helium, on the other hand, is stable, secure, unchanging.

Under heat and pressure two hydrogen become one, and so too your marriage will have heat and pressures, stress and discomfort, but it is under these conditions that the two of you fuse together, that you are sent through the crucible, so to speak, a trail by fire, and you emerge as one element, stronger together, stable, and

secure.

Now, when atoms fuse, another thing

happens, and that of course, is light.

Glorious light, ever-pouring into the

darkness of space. From stars and suns.

Beacons in the night.

And so, two circles, two arrows, pointing at

each other. The convergence of plates

gives us mountains. The fusion of

hydrogen gives us light.

So may you build each other up, and your

love forever shine.

Like a brilliant light on a mountain.

A Toast for Anniversaries

15

You and I, together:

amidst the longest of length

to no length at all.

You and I, together:

amidst the greatest of mass

to no mass at all.

You and I, together:

amidst the largest of current

to no current at all.

You and I, together:

amidst the hottest of heat

to no heat at all.

You and I, together:

amidst the most of units

to no units at all.

You and I, together:

amidst the brightest of lights

to no light at all.

You and I, together:

amidst the passing of time

to no time at all.

You and I, together:

hold fast to me another year,

as I hold fast to you.

*length, mass, electrical current, temperature, amount of a substance, luminous intensity, and time, are the quantities measured by the seven SI base units (meter, kilogram, ampere, kelvin, mole, candela, and second).   All other quantities can be derived from these. Therefore, this toast includes all possible conditions that you and your beloved could experience.

A Blessing for Graduations

16

The maple seed has whirly wings

that spin around and around.

The cottonwood seed with downy fluff

soars softly to the ground.

The coconut seed bobs on the waves,

carried to faraway shores.

The burdock burs with its hooks, stuck

fast, to a slumbering bear as he snores.

Seeds of the dogwood or juniper,

are eaten by birds as a treat,

and carried afar and dropped to the ground

with everything else that birds eat.

So go now, new graduate, into the world,

like a seed, you are ready to fall.

Go now—but don't worry—your time will

come to sprout, take root, and grow tall.

Go now and spin, or fly on the breeze,

or stick to a furry bear,

or fall in the sea, and float across oceans,

or pass through a bird of the air.

In the wind and the rain, the warm or the

frost, cold or sunny days,

go now, like a seed into the world.

Go now, and come what may!

Daily Life

Prayers for Mealtimes

Prayers for Mornings

Prayers for Evenings

Blessings and Goodbyes

Toasts before Coffee, Tea, or Alcohol

Science Proverbs

Prayers for Mealtimes

17

My friends,

We are thankful for this meal,

this gift of food, this gift of life.

We thank those who grew it, shipped it,

purchased it, prepared it,

and we are grateful for the present

company who will eat it.

We are forever amazed by the great chain

of events that has brought us, and this

food, to this table. Here now.

Amen.

18

My friends,

Breathe in... breathe out... breathe in.

Our breath keeps-lit the cellular fire, and

fuel lays before us. We add this fuel to that

fire in gratitude and thanksgiving:

that *it* is, and *we* are, and it *will become* us.

Marvelous matter! Exquisite energy!

Innumerable atoms in multitudinous forms.

Whirling and swirling of subtle force.

Roiling cauldron of delicate perfection. An

orgy of electric nuance — thriving and alive

in such fine range of

temperature and pressure!

This chemical play of life astounds us,

brought to us from the depths of time, and

it will continue within us, with the

ingestion of this food, on and on and into

the future. Amen.

19

My friends,
This food we eat is light —
birthed in the core of our sun, born from
destruction of matter. Fusion of atoms
under pressure — so intense, heat — so
extreme, we cannot fathom nor imagine.
Yet the energy stored in this food has been
there, brought to us by streams of photons
pouring from a star. A single star among
billions. And before that?

All of it, all of them, all of us, traced back
to the singularity of creation.
Mass-energy: in this food, in our minds, in
our fingernails, in billions of galaxies light-
years away, in everything. It is all
connected, from the deepest past, and we
are formed of it.

May we use this light for good thoughts,

good deeds, love, affection, kindness,

compassion; as it passes —on its long

journey —into us, and through us, and out

of us, into a colder cosmos.

Amen.

20

We bless one another,

and these, the gifts,

which we are about to receive,

beloved bounty,

through us and onward,

Amen.

21

My friends,
As we look upon this delicious, flavorful,
wholesome goodness — savory, sweet,
sublime — let us bring to mind those things
that are integral to the digestion of this
meal, responsible for the disassembly of its
fine and complex molecular structure.

We are grateful for saliva, that happy
mixture of enzymes which begin to break
down starch; for the stomach and its
gastric stew, which unravel complex chains
of amino acids. For the intestines and the
billions of bacteria which inhabit them,
symbionts in decay, maestros of chemistry.
We are grateful for the return of this
matter into the great cycles of life, which
flow out of us and into others, and out of

others and back to us. Without these
things we would not be. Too often we
neglect to recognize and appreciate these
crucial processes of disassembly, and our
dependence upon them. We are, after all,
but living beings, which deconstruct other
life to build our own.

We hold these insights in mind as we ingest
this food, and savor its smells, and relish its
flavors, and satisfy our hunger.

Amen.

Prayers for Mornings

22

A new day dawns. I will not waste.

In thought I will reason.

In debate I will use evidence.

In battle I will emerge victorious.

23

Calm, quiet morning.

Slow, silent, still.

Surely it is the Sun that rises.

Surely we are fixed,

and the heavens move above us.

But no, even now we are spinning.

And upright?

There is no upright.

We spin sideways, crossways,

upside down.

And the shaft of light —

growing, rising, rising…?

Nay, we are turning towards it. Ever

turning, round a flaming furnace of fire.

24

What is this living carcass before you?

Limp fingers, resting hands.

You will it — and it moves!

Your eyes, fine-tuned, trace these markings

and your soundless voice springs forth!

Ringing!

But where is it: behind your eyes?

between your ears? in your motionless

mouth?

Thoughts that live nowhere and wander

and vanish…

I will tell you what you are:

You are the sum of uncountable struggles.

You are thousands of generations of

language — words, labels, ideas — passed

into you, which have become you.

Strands of thought compiled from the

ancient past and converged.

You are not alone. How can you be?

The very word 'alone' is a cord to your

human family, to your forebears, to your

friends, to your enemies, to me.

You are the awareness and workings of

trillions of circuits, firing now in crackling

tingling silence.

You will it, and you control matter.

You think it, and your body follows.

You have power over the living carcass

before you. Embody it. You are spirit!

Rise and rise up!

Absorb in rapture your body,

and bend it to your will!

Prayers for Evenings

25

Another day has passed, and the round
earth will roll on while you slumber. It
rolls on without you; it pays no heed of
you; it is but rock and water and air, and
you are but a visitor, an observer, that
neither matters to, nor greatly affects this
massive rolling planet. You are
insignificant to the round earth.
Does this bother you? It should not. You
were born to live and then return, to the
dust, and while you are here, to love your
human family. You need them, and they
need you.
Tomorrow, speak kindly, have patience,
encourage a friend, show interest in

another. The earth is cold and dead but
your brothers and sisters, mothers and
fathers, children, grandchildren, wives and
husbands, friends—they are warm and
alive. They are your home. Take care of
them tomorrow, for tomorrow is another
day. Amen.

26

Now I lay me down to sleep,
The night is lovely, dark and deep.
I will not wake, if I should die,
Perchance to dream, sweet by and by.

Blessings and Goodbyes

27

May the ozone layer shield you,

(from ultraviolet light)

May the magnetic field protect you,

(from the radiating solar wind),

May the heliosphere guard you,

(from high energy cosmic rays)

And may you find peace.

(because it's far more dangerous in space)

28

May your bowels go unobstructed,

your blood flow under reasonable pressure,

and your mitosis be error free.

29

May you grow in the knowledge of science,
may new insights occur to you,
may patterns in data be made apparent
unto you,
and may your theories be ever-verified by
experimental evidence.

30

May you remain forever curious,
uncertain, and wonder-filled.

31

Goodbye my friend. From infinite
possibility, I am me, and you are you, and
we have met, and I am grateful for it.
Go well.

32

I do not wish the wind always at your
back, for then you would not feel it against
your face, and the wind is lovely against
your face.

Toasts before Coffee, Tea, or Alcohol

33

There are no words to improve this drink,

no toast can surpass its virtue —

no invocation or incantation.

It is perfection already.

34

Let us marvel in silence —

At this beloved world, in which friendship

has gathered, and liquids are wonderful.

Cheers!

35

Oh warm beloved roast! Our love affair
continues. On my tongue and within.
Course through me and resurrect my
weary and flaccid flesh, and return this
mind and matter to greater glory!

36

We drink now,
to this suffering life,
eased by liquid hope.

37

Were it not for you, ambrosia,
my thinking mind would be but a curse,
my rational thoughts a despair.

38

Blessed art thou among beans,
and blessed is the brew of thy roast:
coffee!

39

Blessed art thou among yeast, and blessed
is the waste of thy anaerobic metabolism:
alcohol!

Science Proverbs

40

If you're a ribosome, be a ribosome.

If you're a lysosome, be a lysosome,

and be the best lysosome you can be!

Be yourself, be proud of it, and do your job well! The reference here is to both Martin Luther King Jr.'s quote about street sweepers (in which he encouraged street sweepers to sweep streets as Michelangelo painted and as Shakespeare wrote poetry), and to the fact that, in biology classrooms and textbooks for decades, the lysosome has been characterized as the street-sweeper of the cell. The lysosome's job is critical: eliminating and digesting cellular waste. So, no matter what your job, or interests, or niche, or speciality, if you're a lysosome, be a lysosome, and be the best darn lysosome you can be!

41

Don't waste your half-life,
you'll be lead for a long time.

The heaviest natural element that occurs in abundance is uranium, formed by nuclear fusion in supernova explosions. It is unstable and eventually changes, through a series of radioactive decays, to thorium, radium, radon, polonium (all radioactive), and ultimately to a stable isotope of lead. The life-span of a radioactive element is measured in half-life, and once stability is reached, no more decays occur. Although it takes billions of years for the average uranium atom to decay into lead, some of the intermediate elements have much shorter half-lives (radon lasts a matter of days), and lead so very conveniently rhymes with dead, the proverb's lesson is, of course, you'll be dead for eternity, so don't waste your life.

42

Comb your hair,
and make waves the speed of light.

I was utterly fascinated when I was told that by rubbing a comb in my hair (giving the comb a static charge), and then waving it back and forth (accelerating that charge), I was sending light into the world (long wavelength electromagnetic radio waves). Imagine, if you had radio wave vision, the ordinary scene of a woman combing her hair might look as if crackling explosions of light were pouring from her head and comb. This proverb is a reminder that we routinely interact with mysterious, invisible phenomena without realizing it, and you don't need a magic wand or superpowers, you simply need a comb, knowledge of the unseen, and your imagination!

43

Don't jump to translation
until you've finished transcription.

To make a functional protein there is a
specific sequence of events that must be
followed. Nucleic DNA is transcribed into
messenger RNA, that then ventures out of
the nucleus and is translated at ribosomes
into the chains of amino acids that
comprise a protein. DNA that has not
been transcribed completely would result
in mRNA with missing information, which
would be translated into a malformed and
useless protein. So, don't cut corners
before the gun goes off, or rather, don't
jump to translation until you've finished
transcription.

44

Entangled quantum particles are like sausages, it's better not to see them being made.

A reference to "laws are like sausages..." Entanglement is one of the most strange and interesting phenomena known to physics. When particles are entangled, it is as if they don't have properties at all until we observe them. This leads to ideas such as, "observing the universe is what makes it real", or "the moon does not exist when you are not looking at it." The instant you "see" the entangled particles, they lose this bizarre property of not having any properties. Perhaps I should rephrase this proverb to "entangled quantum particles are like sausages, you don't know what's in them."

45

Put one nucleotide in front of the other.

Like feet, step by step, DNA replicating proteins move along the unzipped double helix, and DNA is copied, letter by letter, or nucleotide by nucleotide. Once the DNA is copied, one cell can become two, DNA is copied again, two can become four, DNA is copied again, four can become eight, and on and on, until there you are, trillions of cells strong. And so, to build something great, there is work to do, you just have to put one nucleotide in front of the other. What work do you wish to do? Be like the tireless machinery of the cell, and you will get there.

46

Don't worry, after all,
95% of the universe is "dark".

In the late 1990s, observations of supernova showed that the expansion of the universe was accelerating. This fact has informed our current understanding that 68% of the mass-energy of the universe is "dark energy"—energy from the vacuum of space which drives its expansion. Scientists are unsure how to fully explain this energy. Add to that the 27% of mass that is "dark matter", which has been a mystery since the 1930s, and a full 95% of the mass-energy of the universe is "dark"—the term "dark" being partially a recognition of our ignorance. I find it reassuring to remember, when I am frustrated at my own lack of knowledge and understanding, that the universe continues to surprise us with evidence that forces the revision of our theories and the confession of our ignorance. And so, I try not to worry, after all, 95% of the universe is dark.

47

It's a valence-fill-valence world.

Arguably the most useful concept for explaining and predicting chemical bonding is that of valence. Chemical reactions typically happen such that electrons are given, taken, or shared in order to fill an atom's valence, or outer, shell of electrons. So, this proverb compares atoms not to hungry dogs, competing to fill their bellies at the expense of others, but rather, to cooperative units, negotiating to satisfy one another—giving, taking, and sharing—for increased stability of the whole. It's a valence-fill-valence world, and it's a nice idea.

48

There is more than meets the eye,
and there is more than the eye can see.

Not only are situations frequently more complex than we realize (there are more "factors at play", and much is "hiding under the surface", and there is "more than meets the eye"), but there is also far more than the eye can see: a vast spectrum of electromagnetic radiation of which visible light is but a tiny fraction. So don't be so sure of your conclusions, there is more than meets the eye, and there is more than the eye can see.

49

Sub-sub-sub disciplines of a feather,
flock together.

A proverb with the same meaning as the avian inspired original, but referring to the fact that, with increased knowledge, the tree of science necessarily branches into ever more specific disciplines. The pitfall of increased specialization, however, is that there is perhaps a tendency to "miss the forest for the trees" or here, "miss the flock for the birds" — some conclusion or methodology that could be shared among them, perhaps. There may be unseen connections all around us, perched behind a clump of leaves on the other side of the trunk. Sub-sub-sub disciplines of a feather

flock together, but hopefully we seek connections as they do so.

50

The journey of a thousand serotonin
($C_{10}H_{12}N_2O$) molecules begins with a
single thought.

The neurochemical serotonin and its
associated receptors are implicated in
feelings of happiness, well-being, and
contentment, and so this proverb is a
hybrid of "a journey of a thousand miles
begins with a single step" and "happiness is
contagious". The production of serotonin
and other "happy chemicals" is, of course,

fascinating, yet regardless of the science behind their formation, one can always ask the question: what comes first, a happy thought or a happy feeling? happy mental images, or the release of neurochemicals that make you *feel* happy? Certainly the relationship is one far more complicated than simply this-before-that, yet let us try something:

I want you to imagine.

I want you to imagine a person that you love, a cozy room, a crackling fire. I want you to imagine that person embraces you, holds you tightly. You relax as he or she cradles you and you give yourself to their embrace. You breathe deeply and recline and melt away, with a contented sigh.

Now, read that scene again, slowly, and visualize it, vividly. Really imagine it! Imagine that person, that room, that embrace, that sigh!

Do you feel happier? Perhaps just a little? I would bet that you do. Perhaps my suggestion that you feel happier is what makes you feel like you feel happier. Reader, these are not simply words on a page, these words direct an orchestra of chemicals in your brain, a symphony of thoughts, a fugue of firing neurons, a concerto of ideas, a rhapsody of images, a chorus of connections in the gray thesaurus that is your mind. Your imagination is a most powerful and mysterious phenomena!

The proverb's lesson is this: if you want to feel happy, do the things that make you happy — contentedly happy, serotonin happy. Think happy thoughts, read happy writing, listen to happy people, surround yourself with happy influences, imagine happy things, and be the sower of happiness in others. In effect, the journey of a thousand "happy chemicals" (in you and in those around you), begins with a single thought.

And on a final side note — nay, not a side note but an end note, a finale, an encore — and one which I find particularly ironic and hilarious: the vast majority of serotonin in your body is produced in the intestine and its function is to stimulate intestinal movements. That's right. The same neurochemical that helps you feel contentedly happy keeps your bowels moving, and when released in large quantities in the gut, triggers diarrhea. It's quite true. Does this make you laugh? It makes me laugh. Isn't nature extra-ordinary? Isn't science a blessing?

An Author's Request

My friends,

I hope that you enjoyed this book. I hope that you found at least one blessing that made you laugh, or cry, or think twice, or contemplate; one blessing that you would share with a friend. If you did, then before you go, I would like to ask for your help.

Albert Einstein once said, "try not to become a man of success, but rather a man of value." I believe these science blessings add value to the world. If you agree with me, if you think these blessings are valuable, then help me to spread them. Share them with your family and friends, at gatherings and celebrations. Give them to another science enthusiast, or to a person of faith, or as a present, or in a random act of kindness. Write a review online, with an excerpt of your favorite blessing, and share it through social media, or email, or telephone, or megaphone. You may never know what friend of a friend of a friend will also enjoy this book.

Because, like serotonin molecules, in the brain and in the gut, the journey of a thousand science blessings begins with you. Join me, and in exponential fashion, let us add value to the world. Be a link in the chain that brings these blessings into the lives of others. Share a science blessing, and be blessed in return!

I humbly thank you,

— *J.G. Kemp*

Made in the USA
Monee, IL
18 September 2021

78262873R00062